Plastic Pollution Solutions Guide

Curated by Gary Robinson

This guide is part of the Plastic Pollution Awareness Project
presented by Imaginative Content, LLC
For more information go to
www.PlastiKong.com

© 2026
All Rights Reserved
ISBN 979-8-9939355-2-2
Imaginative Content, LLC
P.O. Box 1123
Santa Ynez, CA 93460
USA

Table of Contents

Introduction

Chapter 1	Understanding Conscious Consumption
Chapter 2	The State of Our Environment
Chapter 3	The Plastic Problem
Chapter 4	Reducing Plastic Waste
Chapter 5	Eco-Friendly Products
Chapter 6	The Power of Policy
Chapter 7	Engaging with Political Leaders
Chapter 8	Educating Consumers
Chapter 9	The Future of Consumption
Chapter 10	Taking Action
Chapter 11	You Can Help End Plastic Pollution
Chapter 12	The Facts About Plastic

Introduction

This little booklet sets out to cover a lot of information in a brief amount of time and space. But the need for both understanding and action are immediate. So few people really grasp the size of the problem or the urgency needed to respond to it.

But no attempt to discuss the reduction of worldwide plastic pollution can begin without both an understanding of the concept of conscious consumption and the current state of the world's environment. Then we can dive into the history of plastic and the enormous size of the plastic problem.

An examination of ideas about reducing plastic waste can then lead to thinking about eco-friendly products. On a grander scale, talking about how politics plays into plastic pollution leads us to realize we need to engage with political leaders.

Educating the public—the worldwide public—about the dangers of plastic waste is at the heart of finding solutions, which will lead to smarter, more environmentally conscious consumers in the future.

With knowledge at our fingertips, in the end, each of us can make an informed choice about how we want to help end plastic pollution.

Note: If you are one of those people who like to skip to the last page or chapter of a book to see how it ends, go ahead and jump to Chapter 11, which dispenses with theories and explanations and lists things you can start doing right now to break your need for plastic and help end plastic pollution.

Chapter 1: Understanding Conscious Consumption

Defining Conscious Consumption

Conscious consumption is a holistic approach to purchasing and using products that considers their environmental, social, and economic impacts. It encourages consumers to be mindful of the choices they make, understanding that these choices can significantly affect the planet and society. This shift in mindset is essential for promoting sustainability and reducing the strain on our natural resources. By prioritizing eco-friendly products, consumers can contribute to a healthier environment and a more equitable economy.

At the heart of conscious consumption is the idea of being informed. Consumers must educate themselves about the origins of products, the materials used, and the ethical practices of companies. This knowledge enables individuals to make choices that align with their values, such as choosing items with minimal packaging or those made from recycled materials. By supporting brands committed to reducing plastic waste and promoting sustainability, consumers can drive demand for environmentally responsible products.

Policy makers and political leaders play a crucial role in fostering an environment conducive to conscious consumption. By implementing regulations that encourage sustainable practices and incentivizing businesses to adopt eco-friendly methods, they can help shift the market toward sustainability. Initiatives such as bans on single-use plastics or incentives for companies that produce biodegradable alternatives can create a landscape where conscious consumption thrives.

The impact of conscious consumption extends beyond individual choices; it can lead to collective action that drives systemic change. When consumers unite in their demand for sustainability, businesses are compelled to adapt or risk losing market share. This cultural shift not only reduces plastic waste but also promotes a circular economy where resources are reused and recycled, minimizing the overall environmental footprint.

In conclusion, defining conscious consumption involves recognizing the power of individual choices in shaping a sustainable future. By prioritizing eco-friendly products and advocating for supportive policies, consumers, policy makers, and political leaders can create a framework for lasting change. Embracing conscious consumption is not just a personal responsibility; it is a collective effort that can lead to a healthier planet for future generations.

The Importance of Sustainable Choices

Sustainable choices are essential in our modern world, where the impact of human activities on the environment is increasingly evident. As consumers, our decisions regarding the products we buy and the companies we support play a significant role in shaping a sustainable future. By opting for eco-friendly products, we encourage businesses to adopt more sustainable practices. This shift in consumer behavior is crucial in combating issues like plastic waste, which poses a significant threat to our ecosystems.

Policy makers and political leaders have a pivotal role in promoting sustainable choices at a broader level. By implementing regulations that favor environmentally friendly products and practices, they can create a framework that incentivizes consumers to make responsible decisions. For instance, policies that support the reduction of single-use plastics can lead to a significant decrease in plastic waste, benefiting both the environment and public health. Engaging with stakeholders, including businesses and environmental organizations, can help in crafting effective policies that resonate with the public.

The concept of conscious consumption extends beyond individual choices; it encompasses a collective responsibility towards our planet. When we prioritize sustainability in our purchasing habits, we contribute to a culture that values ecological preservation over convenience. This change in mindset can lead to innovative solutions and alternatives that address the pressing issues of waste and pollution. By fostering a community that champions sustainable practices, we can drive systemic change that benefits future generations.

Education plays a crucial role in enhancing awareness about the importance of sustainable choices. Consumers equipped with knowledge about the environmental impacts of their purchases are more likely to make informed decisions. Educational campaigns can highlight the benefits of eco-friendly products and the consequences of plastic waste, empowering individuals to take action. Schools, community organizations, and media outlets can collaborate to spread this vital information, creating a society that prioritizes sustainability.

The importance of sustainable choices cannot be overstated. As consumers, policy makers, and political leaders, we must work together to foster an environment where sustainable practices are the norm rather than the exception. By embracing eco-friendly products and advocating for policies that reduce plastic waste, we can make significant strides towards a sustainable world. The choices we make today will shape the future of our planet, and it is our collective responsibility to ensure that it is a healthy and thriving one.

The Role of Consumers in Environmental Impact

Consumers play a pivotal role in shaping the environmental landscape through their purchasing decisions and lifestyle choices. With the growing awareness of climate change and ecological degradation, individuals are increasingly recognizing their power to influence market trends. By opting for eco-friendly products, consumers can push manufacturers to adopt sustainable practices, thereby reducing the overall environmental impact. This shift in consumer behavior not only benefits the planet but also encourages businesses to innovate toward greener solutions.

The impact of consumers is particularly significant in the context of plastic waste reduction. Every year, millions of tons of plastic waste end up in landfills and oceans, causing severe harm to wildlife and ecosystems. By choosing products with minimal or no plastic packaging, consumers can drive demand for alternatives that are biodegradable or recyclable. This conscious decision-making fosters a culture of sustainability and promotes companies that prioritize environmental responsibility.

Moreover, policy makers and political leaders must recognize the influence of consumer behavior in crafting effective environmental policies. By understanding how consumers respond to sustainability initiatives, leaders can develop regulations that incentivize eco-friendly practices. For instance, implementing taxes on single-use plastics or providing subsidies for sustainable products can guide consumers toward making environmentally friendly choices. Such policies not only support the economy but also contribute to a healthier planet.

Additionally, consumer education plays a crucial role in promoting environmental awareness. When consumers are equipped with information about the environmental impacts of their choices, they are more likely to make informed decisions. Educational campaigns can highlight the benefits of sustainable products, encouraging individuals to consider the long-term effects of their consumption habits. As a result, a more informed consumer base can lead to significant changes in market dynamics, pushing industries toward greener alternatives.

The role of consumers in environmental impact is multifaceted and essential for creating a sustainable future. By embracing conscious consumption, individuals can influence corporate practices, drive policy changes, and promote a culture of sustainability. As awareness increases, the collective actions of consumers can pave the way for a healthier environment, demonstrating that every choice matters in the fight against climate change and ecological destruction.

Chapter 2: The State of Our Environment

Current Environmental Challenges

The world is currently facing a multitude of environmental challenges that are intricately linked to consumer behavior and industrial practices. Climate change remains one of the most pressing issues, driven by greenhouse gas emissions from fossil fuels, deforestation, and unsustainable agricultural practices. As consumers, it is imperative to understand how our choices contribute to this global crisis, and to advocate for policies that promote greener alternatives. Every purchase can either exacerbate the problem or support a transition towards a more sustainable future.

Plastic waste represents a significant challenge, with millions of tons of plastic entering our oceans and landfills each year. This pollution not only harms marine life but also poses risks to human health through the consumption of contaminated seafood. Reducing plastic usage requires concerted efforts from consumers, businesses, and policymakers alike. By choosing eco-friendly products and supporting legislation aimed at reducing plastic production, consumers can play a critical role in mitigating this crisis.

Biodiversity loss is also a critical environmental concern, with habitat destruction and pollution leading to the extinction of countless species. The interdependence of ecosystems means that the loss of even a single species can have cascading effects on the environment. As consumers, we can support conservation efforts by making informed choices.

Air and water pollution further complicates the environmental landscape, affecting public health and the planet's ecosystems. Industrial emissions, agricultural runoff, and waste disposal practices contribute to deteriorating air and water quality. Addressing these issues requires a multifaceted approach, including stricter regulations on pollutants and a shift towards cleaner technologies. Consumers can support these initiatives by demanding transparency from companies regarding their environmental practices and by choosing products that have a minimal ecological footprint.

In response to these challenges, a movement towards conscious consumption is gaining momentum. This approach encourages individuals and organizations to make choices that are not only beneficial for their immediate needs but also for the health of the planet. By embracing sustainable practices, and supporting policies aimed at environmental protection, consumers can contribute to a collective effort to combat the pressing environmental challenges we face today.

The Impact of Consumer Behavior on the Planet

Consumer behavior plays a crucial role in shaping the environmental landscape of our planet. As individuals make choices about what to buy, how to use products, and where to dispose of waste, their decisions collectively influence market trends and production practices. Understanding this connection is vital for consumers, policy makers, and political leaders, as it highlights the power of informed consumption in driving sustainable practices and reducing environmental harm.

The rise of eco-friendly products reflects a growing awareness of the impact of consumption on the environment. More consumers are opting for sustainable alternatives, whether it be biodegradable packaging, reusable items, or ethically sourced materials. This shift not only reduces plastic waste but also encourages companies to innovate and adopt greener production methods, ultimately contributing to a healthier planet.

Governments and policy makers have a critical role in facilitating this change. By implementing regulations that support sustainable practices and incentivizing eco-friendly businesses, they can help reshape consumer behavior on a broader scale. Initiatives such as plastic bag bans or incentives for recycling programs can encourage consumers to make more environmentally conscious choices, demonstrating that policy plays a significant role in consumer habits.

Political leaders must also engage in public education campaigns that raise awareness about the consequences of consumer behavior on the environment. By informing the public about issues such as plastic waste and the importance of sustainable consumption, leaders can foster a culture of responsibility and encourage individuals to think critically about their purchasing decisions. This education is essential in promoting long-term behavioral changes that benefit the planet.

Ultimately, the impact of consumer behavior on the planet is profound. As consumers become more aware of their choices and the implications for the environment, their collective actions can drive significant change. By prioritizing sustainability in their consumption habits, individuals can contribute to a more sustainable world, demonstrating that conscious consumption is not just a personal choice but a powerful tool for environmental stewardship.

Global Efforts Toward Sustainability

In recent years, global efforts toward sustainability have gained significant momentum as consumers, policymakers, and political leaders recognize the urgent need to address environmental challenges. Initiatives such as the Paris Agreement demonstrate a collective commitment to reducing greenhouse gas emissions and limiting global warming. These agreements highlight the importance of international cooperation in fostering sustainable practices that can benefit both the planet and its inhabitants.

One prominent area of focus has been the reduction of plastic waste, a critical issue affecting ecosystems and wildlife across the globe. Countries are implementing bans on single-use plastics and promoting the use of biodegradable and eco-friendly alternatives. This shift not only reduces waste but also encourages innovation in product design and materials, leading to a more sustainable economy that prioritizes environmental health.

Consumers are increasingly demanding eco-friendly products, prompting businesses to adapt their strategies and offerings. This shift in consumer behavior is driving companies to invest in sustainable practices, from sourcing materials responsibly to adopting circular economy principles. As a result, the market for green products is expanding, providing consumers with more options that align with their values and the growing emphasis on sustainability.

Policy frameworks play a crucial role in supporting sustainability initiatives. Governments are creating regulations and incentives that encourage businesses to adopt sustainable practices while providing consumers with the tools to make informed choices. By fostering a supportive environment for sustainability, policymakers can drive significant change within industries and promote a culture of conscious consumption among the public.

Ultimately, the path toward a sustainable future requires collaboration among consumers, businesses, and government entities. By working together, these stakeholders can develop solutions that not only address environmental issues but also enhance quality of life for current and future generations. As awareness continues to grow, the collective actions of individuals and organizations around the world will be vital in achieving meaningful progress toward sustainability.

Chapter 3: The Plastic Problem

The History of Plastic Use

The history of plastic use is a fascinating journey that began in the early 20th century. The first synthetic plastic, Bakelite, was created in 1907 by Leo Baekeland, marking a significant turning point in materials science. This invention was celebrated for its versatility and durability, leading to its widespread adoption in various industries, from electrical components to household items. The novelty of plastic captured the imagination of consumers and manufacturers alike, paving the way for further innovations in the field.

As the decades progressed, the development of new types of plastics, such as polyethylene and polyvinyl chloride (PVC), saw an explosion in plastic production and use. By the 1950s, plastic had become a staple in everyday life, used in everything from packaging to toys. The convenience and low cost of plastic products made them immensely popular, but this surge in use began to raise concerns about environmental impacts. The first warnings about plastic waste and pollution started to emerge, highlighting the need for responsible consumption and disposal practices.

The 1970s and 1980s witnessed a growing awareness of plastic pollution, particularly in oceans and waterways. Environmental activists began to advocate for reduced plastic use and more sustainable practices. The introduction of recycling programs aimed to mitigate the increasing plastic waste crisis, but these efforts often fell short due to contamination and lack of infrastructure. This period marked the beginning of a shift in consumer attitudes, as more individuals and organizations started to prioritize eco-friendly alternatives.

In recent years, the conversation surrounding plastic use has intensified, especially with the advent of social media and global movements. The impact of plastic waste on marine life and ecosystems has garnered international attention, leading to policy changes and initiatives aimed at reducing plastic production and consumption. Various countries have implemented bans on single-use plastics, encouraging consumers to seek sustainable options and holding manufacturers accountable for their waste.

Today, as we stand at a crossroads in the history of plastic use, it is crucial for consumers, policy makers, and political leaders to collaborate towards a sustainable future. The lessons learned from the past should guide our choices, emphasizing the importance of conscious consumption and innovation in eco-friendly products. By understanding the history of plastic and its implications, we can work together to reduce plastic waste and protect our planet for future generations.

Environmental Consequences of Plastic Waste

The proliferation of plastic waste has emerged as one of the most pressing environmental challenges of our time. With millions of tons of plastic produced each year, a significant portion ends up in landfills, oceans, and natural habitats. This not only harms wildlife but also disrupts ecosystems, leading to a cascade of negative effects on biodiversity. The durability of plastics means that they can persist in the environment for hundreds of years, breaking down into smaller microplastics that are increasingly found in the food chain, affecting both marine and terrestrial organisms.

One of the most alarming consequences of plastic waste is its impact on marine life. Marine animals often ingest plastic debris, mistaking it for food, which can lead to malnutrition, digestive blockages, and even death. Furthermore, toxins from plastics can leach into the water, accumulating in the tissues of fish and other marine organisms.

This not only poses a threat to marine biodiversity but also raises concerns for human health, as these contaminants can eventually make their way back to our plates through seafood consumption.

Plastic waste also contributes to climate change, as the production and incineration of plastics release significant amounts of greenhouse gases. The extraction of fossil fuels required for plastic production is inherently carbon-intensive, and when plastics are burned, they emit harmful pollutants. Reducing plastic waste is therefore critical in the fight against climate change, necessitating a shift towards more sustainable materials and practices in consumer habits and industrial processes.

Moreover, the economic implications of plastic waste are profound. Communities around the world are grappling with the costs associated with waste management and cleanup efforts. Developing countries are particularly vulnerable, often lacking the infrastructure to deal with plastic pollution effectively. Investment in eco-friendly alternatives and recycling initiatives can create jobs and stimulate local economies, proving that addressing plastic waste is not only an environmental imperative but also an economic opportunity.

To mitigate the environmental consequences of plastic waste, it is crucial for consumers, policymakers, and political leaders to collaborate on effective strategies. Implementing policies that promote the reduction of single-use plastics, incentivizing recycling, and encouraging the use of biodegradable materials are essential steps toward a sustainable future. By making conscious choices and advocating for systemic changes, we can work together to protect our planet for future generations.

The Impact of Microplastics on Ecosystems

Microplastics have emerged as a significant environmental concern, infiltrating ecosystems around the globe. These tiny plastic particles, measuring less than five millimeters, originate from various sources, including the breakdown of larger plastic items and synthetic fibers from clothing. As they accumulate in our oceans, rivers, and soil, they pose severe threats to wildlife and plant life, disrupting natural processes and harming biodiversity. The impact of microplastics is not only ecological but also extends to human health, as they can enter the food chain and potentially affect our well-being.

In marine ecosystems, microplastics are particularly problematic. Marine animals, such as fish and seabirds, often mistake these particles for food, leading to ingestion. This can result in physical harm, such as blockages in the digestive system, and chemical harm due to the toxic substances that microplastics can carry. The consequences of this ingestion ripple through the food web, affecting species at all levels, including those that humans rely on for food. As consumers, awareness of this impact is crucial in driving change towards more sustainable practices.

Freshwater ecosystems are also suffering from microplastic pollution. Rivers and lakes, often seen as pristine natural resources, are not immune to this issue. Studies have shown that microplastics can adversely affect aquatic plants and organisms, disrupting their growth and reproduction. This not only threatens local biodiversity but also compromises the water quality that communities depend on for drinking and agriculture. Policy makers must prioritize the protection of these vital ecosystems to ensure a sustainable future.

Addressing the microplastics crisis requires a multi-faceted approach, emphasizing reduction in plastic use and increased recycling efforts. Consumers play a pivotal role in this equation by making informed choices about the products they purchase, favoring those that minimize plastic packaging and support eco-friendly alternatives.

Furthermore, political leaders must advocate for stringent regulations on plastic production and waste management, as well as invest in research to better understand the impacts of microplastics on ecosystems.

Ultimately, the health of our ecosystems reflects our own health as a society. By recognizing the impact of microplastics and taking collective action, we can work towards a sustainable world where both nature and human communities thrive. Emphasizing the need for conscious consumption can lead to meaningful changes that benefit the environment, promote ecological balance, and safeguard the future for generations to come.

Chapter 4: Reducing Plastic Waste

Strategies for Minimizing Plastic Use

As consumers become increasingly aware of the environmental impact of plastic, strategies for minimizing its use have gained importance. One effective approach is to encourage individuals to adopt a zero-waste lifestyle, which emphasizes reducing, reusing, and recycling items. This lifestyle shift not only cuts down on plastic consumption but also fosters a culture of sustainability that can influence policy and community practices. By prioritizing a zero-waste mindset, consumers can significantly lessen their reliance on single-use plastics.

Another critical strategy involves advocating for policy changes that promote eco-friendly alternatives. Policymakers and political leaders can play a pivotal role by implementing regulations that limit plastic production and incentivize the use of biodegradable materials. For instance, laws that impose taxes on plastic bags or provide subsidies for sustainable packaging options can encourage businesses to transition toward less harmful practices. These policies not only help reduce plastic waste but also create a market for innovative, eco-friendly products.

Education is also essential in the fight against plastic pollution. Awareness campaigns that inform the public about the detrimental effects of plastic waste on the environment can drive home the importance of making conscious choices. Schools, community organizations, and businesses should collaborate to spread knowledge about alternatives to plastic, such as reusable containers and bags. By educating consumers on the benefits of reducing plastic use, communities can build a collective commitment to a sustainable future.

Supporting local and sustainable businesses is another effective strategy. When consumers choose to purchase from companies that prioritize sustainable practices, they help create a demand for eco-friendly products. This shift not only reduces plastic consumption but also strengthens local economies. By opting for products that are packaged with minimal or no plastic, consumers can send a powerful message to larger corporations about the importance of sustainability in their operations.

Finally, community engagement is crucial for fostering a culture of sustainability. Organizing local clean-up events and workshops on reducing plastic waste can empower individuals to take action in their own neighborhoods. These initiatives not only help keep local environments clean but also build a sense of community and shared responsibility. By working together, consumers, policymakers, and political leaders can create a united front against plastic pollution, ultimately leading to healthier ecosystems.

Alternatives to Single-Use Plastics

As the world grapples with the escalating crisis of plastic waste, it is imperative to explore viable alternatives to single-use plastics. These alternatives not only aim to reduce the environmental impact but also provide sustainable options for everyday use. While the convenience of single-use plastics is enticing, innovative solutions are emerging that align with a more eco-conscious lifestyle. From biodegradable materials to reusable products, the options are diverse and increasingly accessible to consumers.

One of the most promising alternatives is the use of biodegradable plastics, which are designed to break down more quickly than traditional plastics. Made from natural materials such as corn starch or sugarcane, these plastics can significantly reduce landfill waste and environmental pollution. When disposed of properly, biodegradable plastics can decompose in a matter of months, offering a viable solution for consumers looking to minimize their plastic footprint.

Another effective alternative is the adoption of reusable products. Items such as stainless steel straws, glass containers, and cloth shopping bags have gained popularity as consumers become more aware of the detrimental effects of single-use plastics. By investing in high-quality reusable options, individuals not only lessen their reliance on single-use plastics but also promote a culture of sustainability. This shift in consumer behavior can have a ripple effect, encouraging businesses to adopt eco-friendly practices and contribute to a more sustainable economy.

In addition to individual choices, policy makers play a crucial role in facilitating the transition away from single-use plastics. Implementing regulations that encourage the use of sustainable materials and taxing plastic production can motivate businesses to seek alternatives. Furthermore, supporting initiatives that promote recycling and waste management will help create an infrastructure that encourages responsible consumption. Collaborative efforts between consumers, businesses, and governments can lead to meaningful change in reducing plastic waste.

Lastly, education and awareness are vital components in the fight against plastic pollution. By informing consumers about the environmental impacts of single-use plastics and the benefits of alternatives, we can foster a more conscious society.

Workshops, community programs, and social media campaigns can serve as platforms for spreading knowledge and encouraging sustainable practices. As consumers become more educated, they are more likely to make choices that align with their values, ultimately leading to a healthier planet.

Community Initiatives for Plastic Waste Reduction

Community initiatives for plastic waste reduction are becoming increasingly vital as the global environmental crisis escalates. Local governments and organizations are stepping up to create programs that encourage individuals and businesses to minimize their plastic consumption. These initiatives often involve educational campaigns that inform residents about the impacts of plastic waste on the environment and human health, fostering a sense of responsibility and collective action.

One effective approach adopted by many communities is the establishment of plastic recycling programs. These programs make it easier for residents to recycle their plastic waste by providing accessible drop-off locations and clear guidelines on what can be recycled. Additionally, some communities are implementing incentives for residents who participate in these recycling efforts, such as discounts at local businesses or recognition in community events, further motivating sustainable practices.

Another significant initiative is the promotion of plastic-free events and markets. By organizing fairs, festivals, and farmers' markets that prioritize eco-friendly products and practices, communities can create spaces where consumers are encouraged to shop without single-use plastics. These events often feature local artisans and vendors who offer sustainable alternatives, helping to shift consumer behaviors and preferences towards more environmentally friendly options.

Collaboration between local governments, businesses, and non-profit organizations is also essential in driving these initiatives forward. By working together, stakeholders can pool resources and expertise, creating comprehensive strategies for plastic waste reduction. This collaborative approach not only amplifies the impact of individual efforts but also fosters a community-wide commitment to sustainability and responsible consumption.

Ultimately, community initiatives for plastic waste reduction serve as a powerful reminder that collective action can lead to significant change. As consumers, policy makers, and political leaders become more engaged in these efforts, the potential for positive environmental impact increases. By supporting and participating in local initiatives, individuals can contribute to ensuring that the planet is preserved for generations to come.

Chapter 5: Eco-Friendly Products

What Makes a Product Eco-Friendly?

Understanding what makes a product eco-friendly is crucial in today's consumer-driven society. Eco-friendly products are designed to have minimal impact on the environment throughout their lifecycle, from production to disposal. This means that the materials used are renewable, sustainable, or recycled, and the manufacturing processes are energy-efficient and free from harmful chemicals. Consumers increasingly demand transparency in the sourcing and production of goods, making it essential for manufacturers to adopt eco-conscious practices.

One of the key factors that contribute to a product's eco-friendliness is its materials. Biodegradable and recyclable materials, such as plant-based plastics or recycled metals, significantly reduce pollution and waste. Additionally, products made from organic materials do not involve harmful pesticides or fertilizers, which can leach into the soil and water supply. By choosing products made from these materials, consumers help reduce the demand for virgin resources and the environmental degradation associated with their extraction.

Energy efficiency is another critical component of eco-friendly products. Items that require less energy during production or use, such as energy-efficient appliances, contribute to lower greenhouse gas emissions. Furthermore, companies that invest in renewable energy sources, like solar or wind power, are making strides toward sustainability. By prioritizing energy efficiency, consumers can choose products that not only save money in the long run but also support a cleaner planet.

Another aspect to consider is the product's packaging. Eco-friendly products often utilize minimal or biodegradable packaging to reduce waste. This is particularly important in the fight against plastic pollution, as traditional plastic packaging can take hundreds of years to decompose. Innovative companies are developing compostable packaging solutions that reduce landfill contributions and promote a circular economy. By supporting brands that prioritize sustainable packaging, consumers can directly impact the reduction of plastic waste.

Finally, a product's end-of-life recyclability plays a significant role in its eco-friendliness. Products designed for easy disassembly and recycling can be reintegrated into the manufacturing cycle, minimizing waste. Consumers should look for items with clear recycling instructions and support initiatives that encourage circular economies. By making informed choices and advocating for sustainable practices, consumers, policymakers, and political leaders can collectively foster a culture of sustainability that benefits the environment and future generations.

Identifying Sustainable Brands

Identifying sustainable brands requires consumers to look beyond marketing slogans and delve into the practices that underpin a company's operations. Brands that prioritize sustainability often have transparent supply chains, ethical labor practices, and a commitment to reducing their environmental footprint. Consumers can start by researching a brand's certifications, such as Fair Trade or Organic, which indicate adherence to specific environmental and social standards. This due diligence empowers consumers to make informed choices that align with their values.

Another crucial aspect is understanding a brand's approach to materials and production processes. Sustainable brands typically utilize eco-friendly materials, such as organic cotton or recycled plastics, and implement processes that minimize waste and energy consumption. By choosing products made from sustainable materials, consumers can significantly reduce their own carbon footprint while supporting companies that prioritize the planet. It's essential to engage with brands that not only claim to be sustainable but also demonstrate responsible practices through their sourcing and manufacturing.

In addition to materials, the longevity and reparability of products are important indicators of sustainability. Brands that design products to last longer and be easily repaired contribute to a circular economy, reducing waste and the demand for new resources. Consumers should look for brands that offer repair services or replacement parts, as this reflects a commitment to sustainability. By investing in durable goods, consumers can minimize their consumption patterns, ultimately leading to a reduction in plastic waste and other environmental impacts.

Furthermore, transparency in corporate social responsibility (CSR) efforts is a key component of identifying sustainable brands. Companies that openly share their sustainability goals, progress, and challenges are more likely to be genuinely committed to their mission. Engaging with a brand's sustainability reports or social media can provide insights into their practices and performance. Consumers should favor brands that actively communicate their efforts to improve and innovate in sustainability, as this fosters accountability and trust.

Lastly, community engagement and support for local initiatives are indicators of a brand's commitment to sustainability. Brands that invest in their communities, whether through partnerships with local environmental organizations or by promoting sustainable practices among their suppliers, demonstrate a holistic approach to sustainability. By supporting such brands, consumers not only make responsible choices but also contribute to broader environmental and social impacts, fostering a more sustainable world for future generations.

The Role of Certifications and Labels

In today's environmentally conscious society, certifications and labels play a pivotal role in guiding consumers toward sustainable choices. These identifiers help consumers identify products that have met specific environmental standards, thereby making it easier to support eco-friendly options. With the increasing awareness of plastic pollution and its impact on our planet, certifications provide a reliable framework for consumers who wish to reduce their ecological footprint through informed purchasing decisions.

Certifications such as Energy Star, Fair Trade, and USDA Organic are more than just marketing tools; they represent a commitment to environmental and social responsibility. For consumers, these labels serve as a shortcut to understanding the sustainability of a product without having to delve deep into its manufacturing process. By choosing products with credible certifications, consumers can feel empowered, knowing they are contributing to a larger movement toward a sustainable future.

Moreover, certifications not only benefit consumers but also encourage manufacturers to adopt more sustainable practices. Companies seeking to obtain these labels often invest in cleaner technologies and sustainable materials, thus reducing their overall environmental impact. This creates a positive feedback loop where consumer demand for certified products drives innovation and accountability in production processes.

Policy makers and political leaders also play a crucial role in the effectiveness of certifications and labels. By supporting legislation that standardizes these certifications, they can ensure that consumers are presented with honest and accurate information about products. This can lead to greater trust in labeling systems, ultimately promoting sustainability on a larger scale and reducing plastic waste in our communities.

In conclusion, the role of certifications and labels in conscious consumption cannot be overstated. They empower consumers to make informed choices, encourage manufacturers to adopt sustainable practices, and create a framework for policy makers to support environmentally friendly initiatives. As we move forward, it is essential that all stakeholders recognize the importance of these certifications in fostering a sustainable world and work collaboratively to enhance their credibility and effectiveness.

Chapter 6: The Power of Policy

Understanding Environmental Policy

Environmental policy serves as the framework for how societies address the pressing issues related to the environment. It encompasses various regulations, laws, and initiatives that aim to protect natural resources and promote sustainability. For consumers, understanding these policies is crucial, as they shape the products available in the market and the practices of businesses. Policy makers and political leaders play a pivotal role in crafting policies that respond to environmental challenges, ensuring that economic growth does not come at the expense of ecological health.

The importance of eco-friendly products cannot be overstated in the context of environmental policy. Policies that encourage sustainable practices often lead to increased demand for eco-friendly alternatives, influencing how consumers make purchasing decisions. This shift not only benefits the planet by reducing waste and pollution but also fosters innovation in product development. As consumers become more aware of environmental issues, their preferences can drive the market towards more sustainable options, prompting companies to adopt greener practices.

Plastic waste reduction is a significant focus within environmental policy, given the global crisis of plastic pollution. Policies aimed at reducing plastic use can include bans on single-use plastics, incentives for recycling, and educational campaigns on waste management. Understanding these initiatives allows consumers to make informed choices that align with their values, while also holding businesses accountable for their environmental impact. As political leaders prioritize plastic waste reduction, they can mobilize resources and create programs that effectively mitigate this pressing issue.

The interaction between consumers and environmental policy is dynamic and essential for fostering a sustainable future. When consumers advocate for stronger environmental regulations, they signal to policy makers the need for change. This collaboration can lead to more robust policies that protect the environment while supporting economic interests. Moreover, the more consumers understand the implications of these policies, the better equipped they are to support initiatives that contribute to sustainability.

In conclusion, understanding environmental policy is vital for consumers, policy makers, and political leaders alike. It empowers individuals to make conscious choices about eco-friendly products and encourages collective action towards plastic waste reduction. By engaging with and influencing environmental policies, all stakeholders can contribute to a healthier planet, ensuring that future generations inherit a sustainable world. The intersection of informed consumerism and effective policy is key to achieving long-term environmental goals.

The Role of Government in Sustainable Practices

Governments play a crucial role in promoting sustainable practices that align with eco-friendly initiatives. By implementing policies that encourage the use of renewable resources and the reduction of plastic waste, they can set a framework that guides both consumers and businesses toward more sustainable choices. Regulations can incentivize companies to innovate, leading to the development of products that are less harmful to the environment and support a circular economy.

Case Studies of Effective Policies

In recent years, various countries have implemented innovative policies to promote sustainability and reduce plastic waste, showcasing effective strategies that can inspire others. One notable case study is the ban on single-use plastic bags in Kenya, which has resulted in a significant decrease in plastic pollution. The government enforced strict penalties for non-compliance, encouraging businesses and consumers to adopt more eco-friendly alternatives. This policy not only addressed environmental concerns but also raised awareness about the detrimental effects of plastic waste on ecosystems and human health.

Another compelling example comes from Sweden, where a comprehensive recycling system has been established. The government incentivizes citizens to sort their waste and rewards them through a deposit return scheme for bottles and cans. This initiative has led to one of the highest recycling rates in the world, demonstrating how effective policy can change consumer behavior and promote a culture of sustainability. The Swedish model illustrates the importance of making recycling accessible and rewarding, encouraging participation at all levels of society.

In Germany, the introduction of the Extended Producer Responsibility (EPR) policy has transformed the way manufacturers handle waste. This policy mandates that producers are responsible for the entire lifecycle of their products, including disposal and recycling. As a result, companies are incentivized to design products that are easier to recycle and have a reduced environmental impact. This approach not only minimizes plastic waste but also fosters innovation in product design, leading to a more sustainable economy.

New Zealand's commitment to a zero-waste goal is another exemplary case study. The government has implemented policies that encourage businesses to reduce waste and promote sustainable practices. By providing grants for eco-friendly business initiatives and setting ambitious waste reduction targets, New Zealand is paving the way for a circular economy. This proactive approach not only benefits the environment but also creates new economic opportunities, showcasing the potential for sustainability to drive growth.

Lastly, the city of San Francisco has set a benchmark with its ambitious waste management policies, aiming for zero waste by 2030. Through mandatory composting and recycling programs, the city has significantly reduced landfill waste. Community engagement and education play crucial roles in this initiative, as residents are encouraged to participate actively in waste reduction efforts. San Francisco's policies illustrate how local governments can take impactful steps toward sustainability, serving as a model for cities worldwide to follow in their quest for a greener future.

Chapter 7: Engaging with Political Leaders

Advocacy and Activism for Sustainable Practices

Advocacy and activism play a crucial role in promoting sustainable practices that benefit both the environment and society at large. Consumers, policy makers, and political leaders must collaborate to create a unified front against environmentally harmful practices. By raising awareness and advocating for change, individuals and organizations can influence policies that support eco-friendly products and reduce plastic waste. Each voice can contribute to a larger movement towards sustainability, emphasizing the importance of collective action.

One of the key aspects of advocacy is education. It is essential to inform consumers about the environmental impact of their choices and the benefits of sustainable products. Workshops, seminars, and social media campaigns can empower individuals to make informed decisions that align with their values. By understanding the consequences of plastic waste and the advantages of eco-friendly alternatives, consumers can drive demand for products that prioritize sustainability.

Policy makers play a vital role in shaping the landscape for sustainable practices. They have the power to implement regulations that encourage businesses to adopt environmentally friendly methods. Advocacy efforts aimed at policy reform can lead to significant changes, such as bans on single-use plastics or incentives for companies that prioritize sustainability. Engaging with political leaders to promote these initiatives can create a ripple effect, influencing broader societal changes.

Activism also involves holding corporations accountable for their environmental impact. Consumers are increasingly demanding transparency and ethical practices from brands. By supporting companies that prioritize sustainability and protesting against those that do not, individuals can drive the market towards more responsible practices. Grassroots movements and partnerships with environmental organizations can amplify these efforts, ensuring that the message of sustainability reaches a wider audience.

Advocacy and activism are essential components in the pursuit of sustainable practices. By fostering collaboration among consumers, policy makers, and political leaders, we can create a more sustainable world. The collective efforts of individuals and organizations can lead to significant change, reducing plastic waste and promoting eco-friendly products. Together, we can ensure that future generations inherit a healthier planet.

Building Relationships with Policymakers

Building relationships with policymakers is essential for fostering a sustainable future. Consumers, political leaders, and environmental advocates must come together to create dialogues that prioritize eco-friendly initiatives. By collaborating with policymakers, individuals can influence legislation that supports the reduction of plastic waste and promotes environmentally responsible practices.

Effective communication is key in building these relationships. Consumers can share their concerns regarding environmental issues with policymakers through various channels, such as social media, public forums, or direct meetings. This engagement not only raises awareness but also provides policymakers with the insights needed to develop policies that reflect the public's values and desires for a sustainable world.

In addition to communication, establishing trust is crucial. Policymakers must see consumers as partners rather than adversaries. This can be achieved by demonstrating a genuine commitment to sustainability through collective actions, such as participating in community clean-ups or supporting local eco-friendly businesses. When policymakers recognize the dedication of consumers, they are more likely to prioritize environmental issues in their agendas.

Furthermore, educating policymakers on the impact of plastic waste and the benefits of eco-friendly products is vital. Hosting workshops, webinars, and informational sessions can equip political leaders with the knowledge to make informed decisions. By showcasing successful case studies and innovative solutions, consumers can illustrate the positive outcomes of sustainable practices, reinforcing the importance of their incorporation into policy frameworks.

Finally, maintaining ongoing relationships with policymakers ensures that environmental issues remain on their radar. Regular follow-ups, updates on consumer sentiments, and continued advocacy can keep the momentum alive. By fostering these connections, consumers can help shape a legislative landscape that champions sustainability, ultimately leading to a healthier planet for future generations.

The Importance of Grassroots Movements

Grassroots movements play a crucial role in shaping policies and attitudes toward sustainability. These community-driven initiatives often emerge from a collective desire to address local environmental issues, such as plastic waste and the promotion of eco-friendly products. By mobilizing individuals at the local level, grassroots movements can create significant pressure on policymakers to implement necessary changes that benefit the environment and society as a whole.

One of the key strengths of grassroots movements is their ability to raise awareness and educate the public about environmental challenges. Through campaigns, workshops, and community events, these movements empower consumers to make informed choices. This education is essential in promoting the adoption of sustainable practices, as it encourages individuals to reconsider their consumption habits and opt for products that align with their values.

Moreover, grassroots movements foster a sense of community and collective action. When individuals come together for a common purpose, they can drive impactful change more effectively than isolated efforts. This collaboration not only strengthens community bonds but also amplifies their voice in advocating for policies that promote sustainability. By uniting around shared goals, grassroots movements can challenge corporations and governments to prioritize eco-friendly solutions.

In addition to raising awareness, grassroots movements often serve as incubators for innovative solutions to environmental problems. They encourage creativity and experimentation, allowing new ideas to flourish in a supportive environment. For instance, local initiatives may develop unique recycling programs or community gardens that showcase sustainable practices and inspire others to follow suit, demonstrating the potential for change when individuals take the initiative.

Finally, the influence of grassroots movements extends beyond local communities, shaping national and even global conversations about sustainability. As these movements gain traction, they can influence larger organizations and policymakers to adopt more comprehensive environmental policies. Ultimately, the importance of grassroots movements lies in their ability to catalyze change, empower consumers, and create a more sustainable world for future generations.

Chapter 8: Educating Consumers

The Role of Education in Conscious Consumption

Education plays a pivotal role in shaping the mindset of consumers towards conscious consumption. Through educational initiatives, individuals are equipped with the knowledge to make informed decisions about their purchases. Understanding the environmental impact of products encourages consumers to opt for eco-friendly alternatives, reducing their carbon footprint and supporting sustainable practices. This awareness is crucial in combating issues like plastic waste, which has reached alarming levels globally.

In schools and communities, programs focused on sustainability can foster a culture of responsibility among young consumers. By integrating lessons about the lifecycle of products and the importance of reducing plastic usage into the curriculum, educators can inspire the next generation to prioritize environmental considerations in their choices. This foundational knowledge empowers students to advocate for change and become informed consumers who value sustainability.

Moreover, education extends beyond the classroom and into public awareness campaigns that engage consumers of all ages. These campaigns can highlight the consequences of overconsumption and the benefits of choosing eco-friendly products. Policy makers and political leaders can also play a significant role by supporting these initiatives and ensuring that the message of conscious consumption reaches every corner of society.

In addition to traditional education, digital platforms provide an avenue for widespread knowledge dissemination. Online courses, webinars, and social media campaigns can reach diverse populations, promoting sustainable practices and conscious consumption. By leveraging technology, educators and advocates can create a more informed public that is aware of their consumption habits and the environmental impact they hold.

Ultimately, the synergy between education and conscious consumption has the potential to reshape consumer behavior. As individuals become more knowledgeable about the implications of their choices, they are likely to support policies that align with sustainability goals. This collective shift towards informed consumption can lead to a significant reduction in plastic waste and a healthier planet for future generations.

Resources for Informed Decision-Making

In the age of conscious consumption, accessing reliable resources is crucial for informed decision-making. Consumers, policymakers, and political leaders alike must navigate a complex landscape filled with information about sustainability, eco-friendly products, and plastic waste reduction. Utilizing credible sources enhances understanding and promotes choices that positively impact the environment. This subchapter will explore various resources available for individuals and groups committed to making sustainable decisions.

One of the most valuable resources for informed decision-making is comprehensive databases and websites dedicated to environmental issues. Organizations such as Greenpeace and the UN's Environment Programme provide extensive research and data on sustainability practices and eco-friendly products. These platforms often feature guides, tools, and reports that can empower consumers to recognize environmentally harmful practices and make better choices. Additionally, they offer insights into the lifecycle of products, helping individuals understand the implications of their purchases on plastic waste.

Educational initiatives and workshops also serve as excellent resources for consumers and leaders seeking to deepen their understanding of sustainable practices. Local governments and community organizations frequently host events that focus on eco-friendly living and waste reduction strategies. Participating in these initiatives not only provides knowledge but also fosters community engagement and collaboration. By sharing experiences and best practices, attendees can inspire one another to adopt more conscious consumption habits.

Furthermore, social media platforms and online forums have emerged as vital resources for sharing information and experiences related to sustainability. Influencers and activists dedicated to environmental causes often use these platforms to educate their followers about eco-friendly products and practices. Engaging with these communities enables consumers to exchange ideas and solutions to challenges faced in reducing plastic waste.

Lastly, publications such as books, articles, and journals focused on sustainability provide in-depth analysis and critique of current practices. These resources can be instrumental for policymakers and political leaders as they formulate strategies to combat environmental issues. By referencing scholarly work and case studies, decision-makers can craft informed policies that support sustainable development. Ultimately, harnessing a variety of resources empowers all stakeholders to make knowledgeable choices that contribute to a sustainable world.

The Impact of Social Media on Consumer Choices

Social media has transformed the way consumers make choices, particularly regarding environmentally friendly products. With platforms such as Instagram and TikTok, brands have the ability to showcase their eco-friendly initiatives and products to a global audience. This exposure has led consumers to become more aware of the impact their choices have on the environment, encouraging a shift toward sustainable consumption practices. The ability to share experiences and reviews online also influences purchasing decisions, as consumers increasingly trust peer recommendations over traditional advertising.

The role of influencers in shaping consumer preferences cannot be overstated. Many eco-conscious influencers promote sustainable living and highlight brands that prioritize environmental responsibility. Their reach and credibility can significantly affect the choices made by their followers, leading to increased demand for products that minimize plastic waste. This trend not only benefits consumers seeking greener options but also pressures companies to adapt their practices to meet this rising demand.

Furthermore, social media serves as a platform for raising awareness about pressing environmental issues. Campaigns highlighting the dangers of plastic pollution and the importance of recycling have gained traction, often going viral and mobilizing large audiences. These movements encourage consumers to rethink their consumption habits, prompting them to choose products that align with their values. The ability to rally support and create community around sustainability initiatives illustrates the powerful impact social media can have on consumer behavior.

However, while social media promotes positive changes, it also has its downsides. Misinformation about products and practices can spread rapidly, leading consumers to make misguided choices. Not all brands that claim to be eco-friendly are genuinely sustainable, and consumers must navigate these claims carefully. This highlights the need for increased transparency and education around sustainable practices, ensuring that consumers can make informed decisions.

In conclusion, the interplay between social media and consumer choices is complex and multifaceted. As consumers, policymakers, and political leaders navigate this evolving landscape, it is crucial to harness the power of social media to promote sustainable products and practices effectively. By fostering an informed community and encouraging responsible consumption, social media can be a force for good in the journey toward a more sustainable world.

Chapter 9: The Future of Consumption

Trends in Sustainability

In recent years, the focus on sustainability has surged, driven by an increasing awareness of environmental issues and the urgent need for action. Consumers are becoming more conscious of their purchasing choices, seeking eco-friendly products that minimize harm to the planet. This shift is not just a trend but a reflection of a deeper understanding of the interconnectedness of human activities and the health of our ecosystems. As a result, brands are innovating to meet this demand, leading to a marketplace that prioritizes sustainability in its offerings.

Policy makers are also responding to the call for sustainability through legislation and initiatives aimed at reducing plastic waste and promoting environmentally friendly practices. These efforts range from implementing stricter regulations on single-use plastics to providing incentives for businesses that adopt sustainable practices. Political leaders play a crucial role in shaping the framework within which these changes occur, emphasizing the importance of collaboration between government, industry, and consumers to create a sustainable future.

The rise of sustainable practices is evident in various sectors, from food and fashion to technology and construction. Companies are investing in renewable energy sources and sustainable materials, reducing their carbon footprints while appealing to a growing base of eco-conscious consumers. This trend not only benefits the environment but also enhances brand loyalty, as consumers are more likely to support businesses that align with their values of sustainability and responsibility.

Educational initiatives are critical in fostering a culture of sustainability. As consumers become more informed about the impact of their choices, they are empowered to demand accountability from brands and policy makers. Schools and organizations are increasingly emphasizing environmental education, helping to cultivate a generation that prioritizes sustainability in both personal and professional spheres. This knowledge is essential for driving the systemic changes needed to address the challenges posed by climate change and environmental degradation.

Ultimately, the trends in sustainability reflect a collective shift towards a more conscious and responsible way of living. As consumers, policy makers, and political leaders collaborate to promote eco-friendly practices and reduce plastic waste, the potential for meaningful change increases. The future of sustainability lies in our ability to work together, innovate, and commit to making choices that benefit not just ourselves but the planet as a whole.

Innovations in Eco-Friendly Products

Innovations in eco-friendly products have gained significant momentum as consumers become more aware of their impact on the environment. Companies are now focusing on sustainable materials and production methods that minimize plastic waste while offering functional and appealing alternatives. This shift not only addresses environmental concerns but also aligns with the growing consumer demand for responsible choices that reflect their values.

One of the most notable innovations is the development of biodegradable materials that can replace traditional plastics. These materials break down more easily in the environment, reducing the time they contribute to pollution. For example, products made from plant-based polymers are gaining traction, providing consumers with familiar products that do not harm the planet in the same way conventional plastics do.

Additionally, many companies are exploring circular economy models, where products are designed for reuse and recycling. This approach encourages consumers to return used items for repurposing, thus extending the lifecycle of materials and minimizing waste. By investing in such innovative business models, organizations not only cater to eco-conscious consumers but also promote a sustainable economy that benefits all.

Government policies play a crucial role in fostering these innovations by incentivizing research and development in eco-friendly technologies. Political leaders can create frameworks that support businesses in adopting sustainable practices, which in turn encourages more companies to innovate. By establishing standards and regulations that prioritize environmental health, policymakers can drive the market towards greener alternatives.

The push for eco-friendly products represents a significant step towards reducing plastic waste and promoting sustainability. Consumers, policy makers, and political leaders all have a vital role to play in this transformation. By embracing innovations in eco-friendly products, we can collectively contribute to a healthier planet and a more sustainable future.

The Vision for a Sustainable World

The vision for a sustainable world is rooted in a fundamental shift in how we, as consumers and leaders, perceive our relationship with the environment. It involves recognizing that every choice we make has a ripple effect on our planet and its resources. By embracing conscious consumption, we can collectively work towards a future where ecological balance and human well-being are prioritized. This vision is not just about reducing our carbon footprint; it's about creating a thriving ecosystem for generations to come.

A key aspect of this vision is the promotion of eco-friendly products that minimize environmental impact. These products not only reduce waste but also encourage sustainable practices among consumers. By choosing items made from renewable resources, biodegradable materials, or those that support circular economies, we can help shift the market towards more responsible production methods. This transition is essential for reducing plastic waste, which poses a significant threat to our oceans and wildlife.

Policy makers play a crucial role in realizing the vision of a sustainable world. They have the power to implement regulations that encourage sustainable practices and discourage harmful ones. By instituting policies that support renewable energy, incentivize recycling programs, and impose stricter limits on plastic production, policy makers can drive significant change. This collaboration between consumers and leaders is vital in building a framework that fosters sustainability at every level of society.

Political leaders must also champion this vision by aligning their agendas with sustainable development goals. They can advocate for environmental justice and prioritize initiatives that address climate change and pollution. By engaging communities in discussions about sustainability, leaders can empower individuals to take actionable steps towards reducing their environmental impact. This grassroots movement, fueled by informed consumers and supportive policies, can lead to widespread changes in behavior and attitudes.

Ultimately, the vision for a sustainable world is achievable if we collectively commit to making conscious choices. It requires a shift in mindset from convenience to responsibility, where consumers, policy makers, and political leaders work in unison. By fostering a culture of sustainability, we can ensure that our planet remains a vibrant and healthy place for future generations, free from the burdens of plastic waste and environmental degradation.

Chapter 10: Taking Action

Steps for Individual Change

Individual change is a powerful catalyst for broader societal transformation, especially in the context of sustainable living. Consumers, policy makers, and political leaders must recognize that their choices, whether conscious or unconscious, significantly impact the environment. By committing to eco-friendly practices, individuals can drive demand for sustainable products and influence market trends, encouraging companies to adopt greener practices.

The first step towards individual change is awareness. It is essential for consumers to educate themselves about the environmental consequences of their choices, particularly regarding plastic waste and the products they consume. Understanding the lifecycle of products, from production to disposal, can help individuals make informed decisions that align with their values and environmental goals.

Next, consumers should evaluate their purchasing habits critically. This involves questioning the necessity of each product and opting for eco-friendly alternatives whenever possible. Simple actions, such as choosing reusable bags, bottles, and containers, can significantly reduce plastic waste. Additionally, supporting businesses that prioritize sustainability can amplify the impact of individual choices and encourage more companies to follow suit.

Policy makers and political leaders play a crucial role in facilitating individual change by creating supportive frameworks. This includes implementing regulations that promote sustainable practices and providing incentives for consumers to choose eco-friendly options. By fostering an environment where sustainable choices are accessible and affordable, leaders can empower citizens to adopt more responsible consumption habits.

Finally, collective action amplifies individual efforts. When consumers unite around sustainability initiatives, they can create a powerful movement that influences corporate policies and government regulations. Participating in community clean-up events, advocating for policies that reduce plastic waste, and sharing knowledge about sustainable practices can inspire others to join the cause, leading to a more sustainable world for all.

Collaborating for Collective Impact

In today's complex world, the challenges of environmental degradation and plastic waste require collaborative efforts to achieve meaningful change. Consumers, policymakers, and political leaders must come together, recognizing their unique roles and responsibilities in crafting solutions that foster sustainable practices. By forming partnerships across various sectors, they can leverage their collective influence to promote eco-friendly products and reduce plastic waste, leading to a healthier planet for future generations.

Collaboration begins with understanding that individual actions, while important, are often insufficient to drive systemic change. Consumers play a crucial role by supporting businesses that prioritize sustainability, while policymakers can create regulations that encourage environmentally friendly practices. Political leaders must champion these initiatives, ensuring that they remain on the agenda and receive the necessary funding and support. This synergistic approach amplifies impact and drives innovation in sustainable consumption.

Building a coalition of stakeholders allows for the sharing of resources, knowledge, and best practices. For instance, businesses can collaborate with environmental organizations to develop eco-friendly alternatives to plastic products, while consumers can provide valuable feedback on these innovations. Such partnerships foster a culture of accountability and transparency, where all parties are motivated to adhere to sustainable practices and contribute to a collective impact.

Furthermore, effective communication is vital in these collaborations. Raising awareness about the benefits of sustainable consumption and the importance of reducing plastic waste can galvanize public support and encourage broader participation. Campaigns that highlight successful collaborations can inspire others to join the movement, demonstrating that collective action can lead to significant environmental improvements.

Ultimately, the success of these collaborative efforts hinges on a shared vision for a sustainable future. By aligning goals and working towards common objectives, consumers, policymakers, and political leaders can create a powerful force for change. The journey towards conscious consumption and a reduction in plastic waste is a collective one, and through collaboration, we can pave the way for a more sustainable world.

Inspiring Others to Make Sustainable Choices

In today's world, the urgency for sustainable choices has never been clearer. Consumers, policymakers, and political leaders play crucial roles in driving the shift towards eco-friendly practices. By inspiring others to embrace sustainable living, we can collectively reduce our impact on the environment and promote a healthier planet. The power of individual choices, when multiplied across communities, can lead to significant change, particularly in areas like plastic waste reduction.

One effective way to inspire sustainable choices is through education and awareness. By providing information on the environmental consequences of our purchasing decisions, we can empower individuals to make informed choices. Workshops, community events, and digital campaigns can serve as platforms to share knowledge about eco-friendly products and the importance of reducing plastic waste. When people understand the direct impact of their actions, they are more likely to adopt sustainable habits.

Moreover, showcasing success stories can motivate others to follow suit. Highlighting individuals, businesses, or communities that have successfully implemented sustainable practices can create a ripple effect. These narratives can illustrate the tangible benefits of conscious consumption, such as cost savings, health improvements, and enhanced community well-being. By celebrating these achievements, we can encourage more people to join the movement towards sustainability.

Policy initiatives also play a pivotal role in fostering sustainable choices. Political leaders have the power to implement regulations that promote eco-friendly practices and discourage wasteful behaviors. Incentives for businesses that prioritize sustainability can stimulate innovation and make green products more accessible to consumers. By aligning policy with the values of sustainability, leaders can create an environment where making eco-friendly choices becomes the norm rather than the exception.

Ultimately, inspiring others to make sustainable choices requires a collaborative approach. Consumers, policymakers, and political leaders must work together to create a culture of sustainability. By sharing knowledge, celebrating successes, and implementing supportive policies, we can pave the way for a more sustainable future. The journey toward conscious consumption is not just an individual endeavor; it is a collective mission that calls for action from all sectors of society.

Chapter 11: You Can Help End Plastic Pollution

We all can significantly help reduce plastic pollution by making conscious choices to minimize plastic consumption and properly manage the plastic we do use. If you like to skip to the end of a book to find out how it ends, here is the meat of the matter:

1. Reduce:
- Minimize single-use plastics: Say no to plastic water and soda bottles, bags, straws, utensils, and takeout containers whenever possible.
- Choose reusable options: Carry a reusable water bottle, coffee cup, shopping bags, and utensils.
- Buy in bulk: This reduces packaging waste.
- Avoid individually wrapped items: Opt for products with less packaging.
- Consider alternatives to plastic: Choose glass, metal, or bamboo products.
- Cook more: Prepare meals at home to reduce reliance on takeout containers.

2. Reuse:
- Repurpose items: Give plastic containers a second life as storage or organization tools.
- Repair and maintain plastic items: Extend the life of your plastic products to reduce the need for replacements.
- Donate or sell used plastics: Give unwanted plastic items a new home.

3. Recycle:
- Recycle properly: Ensure you're using your local recycling programs effectively.
- Support businesses with sustainable packaging: Choose products from companies that prioritize eco-friendly packaging.

4. Advocate:
- Support policies that address plastic pollution: Advocate for plastic bag bans or taxes, and other initiatives.
- Educate others: Share information and encourage others to make changes.
- Participate in cleanups: Help remove plastic waste from the environment.

Political Action - Support policies to reduce single-use plastic.

Imagine being able to buy and use everything your family needs without sending anything to the landfill. That's the vision of a circular economy. This idea is to "design out" waste by switching to products and materials that are meant to stay in use for as long as possible, then be recovered and regenerated into new products.

To make this happen, manufacturers and waste managers must work together to improve both product packaging and waste disposal. That includes making sure that recyclable materials are actually recycled, and that compostable materials are actually composted.

Government policies that restrict certain types of single-use plastic in favor of reusable alternatives help accelerate the transition away from single-use plastic and toward more environmentally responsible alternatives.

Here's what you can do:

- <u>Sign</u> up online for alerts and emails from environmental organizations so you can be notified about opportunities to take action.

- By supporting policies to reduce plastic, you are telling policymakers that ocean health is important to you.

- You can also urge your elected officials to improve waste management systems, including recycling.

MORE WAYS TO DECREASE PLASTIC POLLUTION

Every minute, one garbage truck worth of plastic is dumped into our oceans, and the consequences are astounding. By 2050, there will be <u>more plastic than fish</u> in the ocean by weight. More than 1 million marine animals die each year due to plastic debris.

Plastic leaches toxic chemicals into our environment and food chain, exposing humans to harmful endocrine disruptors. And recently, a <u>deep submersible dive</u> made the horrific discovery of a plastic bag floating near the bottom of the Mariana Trench: Even the deepest part of the ocean, nearly 7 miles down, isn't safe from plastic pollution.

We can all live with less plastic. Start small. Choose a few simple, manageable tips from this list and then build up toward a more plastic-free lifestyle. After all, many people doing it imperfectly is better than a few people doing it perfectly.

1. Decline straws at restaurants and drive-throughs. Use a reusable stainless steel or glass straw instead.

2. Buy popcorn kernels from bulk bins (using a glass jar) and pop them in a Dutch oven instead of using bags.

3. Use a countertop compost bin for food scraps to minimize garbage bag use.

4. Invest in reusable silicone zip-top sandwich bags.

5. One billion plastic toothbrushes are thrown out annually. Use sustainable and biodegradable bamboo toothbrushes.

6. Cut down on excess packaging by buying grains, nuts, legumes, baking ingredients, cereals, and more in bulk.

7. Choose produce with no packaging or with biodegradable wrappers like banana leaves.

8. Use a reusable produce bag made of biodegradable material like cotton, hemp, jute, or bamboo.

9. Plastic bags make up 11.18% of plastic pollution. Support local, regional, and nationwide legislation to ban plastic bags.

10. Bring plastic bags to recycling drop-off locations. You can find one at PlasticFilmRecycling.org.

11. Swap plastic wrap for a reusable wrap made from beeswax or cotton. You can also save food in glass storage containers

12. The DoneGood app can help you find businesses and brands committed to sustainable practices.

13. Single-use coffee pods take from 150 to 500 years to break down. Use reusable pods that you fill with ground coffee instead.

14. Decline getting a receipt whenever possible. Some are coated in a thin layer of plastic.

15. Enjoy ice cream in a cone, not a cup.

16. Buy used plastic items and refurbished electronics.

17. When ordering delivery, tell the restaurant you don't need plastic cutlery.

18. When ordering pizza, request that they do not put in the little plastic table, or "pizza saver."

19. Use reusable shopping bags. Post reminders around your home and car so you don't forget to bring them!
20. Use vinegar (from a glass container) and water for cleaning.
21. Use baking soda that comes in a cardboard box for scrubbing.
22. If you have a sweeper-mop that uses disposable cleaning pads, use reusable pads made from cloth instead.
23. Try to avoid any personal care products with polyethylene listed as an ingredient.
24. Bottles and bottle caps make up 15.5% of plastic pollution. Ditch bottled water: Use a stainless steel, glass, or bamboo water bottle.
25. Switch to plastic-free chewing gum. Most gums are gummy because they are made from plastics, rubbers, and waxes.
26. Bring a reusable food storage container to restaurants for leftovers.
27. Use the app CORKwatch to determine if a wine uses a plastic or natural cork.
28. DON'T litter. Just don't do it! Volunteer at local nature cleanup events.
29. Food wrappers and containers make up 31.14% of plastic pollution by unit count. Try to go a day without buying anything with plastic packaging. Then, try a week. Then, a month. The experience will be eye-opening!
30. Bring your own cup or tumbler to coffee shops. Many places will even give you a discount!
31. Use cotton swabs with paper rods instead of plastic.
32. When you must buy plastic, choose clear plastic bottles (for cosmetics, foods, toys, etc.), which are more likely to be recycled.
33. Don't release balloons or plastic confetti into the air. Try not to purchase or use balloons at all. Confetti made from fallen leaves is a better option!
34. Plastic flip-flops are forever: Choose rubber, cork, jute, or recycled footwear instead.
35. Try bar shampoos and soaps without packaging, like the ones you can buy at Lush.
36. Buy bread from bakeries that package in paper.

37. Properly secure your garbage bags to prevent fly-away plastics while being transported by garbage trucks.

38. Cut down on junk mail (often full of plastic).

39. Don't buy clothing made from synthetic material like polyester, acrylic, Lycra, spandex, or nylon. Cotton, linen, and hemp are better material options.

40. Up to 20 billion pads, tampons, and applicators are dumped into North American landfills annually. Look for alternatives instead.

41. Shop at your local farmers' market and bring your own bags and containers. They often use way less plastic packaging.

42. Avoid buying new CDs and DVDs. Stream or buy used.

43. Crayola recycles all markers. Set up a ColorCycle program through a local school.

44. Staples recycles electronics, ink, and batteries. In Canada, Staples has partnered with TerraCycle to recycle writing utensils, too

45. Use a razor with replaceable blades instead of disposables.

46. Using packing peanuts made from starch instead of Styrofoam.

47. Cutting up plastic six-pack rings still harms marine life, since they can ingest the smaller pieces. Avoid them and support companies that are developing alternatives.

48. Cigarette butts (made from cellulose acetate, a type of plastic) are the largest source of single-use plastic pollution. Kick the habit!

49. The most toxic plastics are #3 (PVC), #6 (polystyrene), and #7 (other, including BPA). Avoid them.

50. Develop your own ways to reduce plastic pollution!

GOOD RULE OF THUMB:

Before buying or using plastic, imagine it in a landfill or in the ocean forever. Taking a moment to reflect on the consequences can compel you to find a non-plastic solution.

Chapter 12: The Facts About Plastic

Scientists and environmentalists have been studying plastic and its effects on the planet for fifty years. So, the results are in. There is no debate. There is no excuse! Oil-derived plastic is bad for people, bad for the planet! Here are some of the facts.

In 2025, a prominent medical journal (Lancit Medical Journal) proclaimed that plastics are a "grave, growing and under-recognized danger" to people and the planet that is causing "disease and death" in humans from infancy to old age.

The headline for this story, published in April of that year, stated that the "Plastics Crisis Costs Trillions, Kills Hundreds of Thousands Each Year." A very bold and astounding statement.

The article goes on to tell us that "plastics are not as inexpensive as they appear and are responsible for massive hidden economic costs borne by governments and societies. These impacts fall disproportionately upon low income and at-risk populations."

But these statements are not based on vague generalities. They are based on the focused review of multiple verified international studies. In other words: science. Not just American science or European science. Worldwide science.

For example, plastics have been found to negatively affect human health in all phases of production, from extracting the fossil fuels that make up 98% of plastics, to use and disposal. One chemical alone, known as BPA, was connected to more than 5 million cases of heart disease and almost 350,000 cases of stroke in one year. That chemical is one of more than 16,000 chemicals present in some plastics.

Another example includes an estimated 158,000 premature deaths worldwide and health-related economic losses in the range of $200 million. Wow!

What quickly becomes very clear from these studies is that these problems will not go away on their own. They will get worse. Therefore, bold, decisive steps must be taken to curtail plastic production, distribution and use, and these steps are needed RIGHT NOW!

When it comes to regulating the production of plastic, part of the problem is transparency. Chemical companies can play "hide the data" and/or "camouflage the data" when revealing the ingredients of the plastic they're producing. If one poisonous chemical is banned, then another similar ingredient with a different name can be created and inserted in the formula. Plastic chemicals come into being faster than they can be regulated.

The Lancit article states, "Despite their large production volumes and widespread human exposure, hazard information remains missing for thousands of chemicals in everyday use."

Why are we so focused on this one article in the Lancit Medical Journal? Because it provides the most comprehensive, definitive collection of information about the dangers, impacts, and absolute horrors of plastic! Hundreds of websites and thousands of individuals have warned of these very real outcomes. And yet, corporations continue to pollute us and out world with impunity while governments ignore or outright support these activities.

We must each find a way to get involved in solving this massive problem. We are the solution to plastic pollution. Activate! Donate! Participate.

Sources of Information for this Guide:

https://myplasticfreelife.com/plasticfreeguide

https://www.earthday.org/2018/04/05/fact-sheet-plastics-in-the-ocean/

https://oceancrusaders.org/plastic-crusades/plastic-statistics

https://conserveturtles.org/information-sea-turtles-threats-marine-debri/

https://www.terracycle.com/en-CA/

https://www.breastcancer.org/risk/factors/plastic

https://plastic-pollution.org

https://www.5gyres.org/plastic-pollution-facts

https://greenpeace.org/international

https://plasticpollutioncoalition.org

https://www.unep.org/

https://www.breakfreefromplastic.org/

The Plastic Pollution Awareness Project
Presented by Imaginative Content, LLC
© 2025 All Rights Reserved

Our world is drowning in plastic. Plastic pollution is *the* environmental crisis of our time. Plastic waste in its many forms now threatens all life on the planet.* Microplastics and Nano-plastics are in the soil, the water, the rain, and the air. These minute particles have been discovered in all species of wildlife, in human bodies, and in our brains! Because the negative effects of this pollution are invisible or inconsequential to most people, we must develop new, innovative ways to get the word out about this issue. We must communicate the message through media that **treat plastic like the dangerous monster it is!** Introducing The Plasti Kong® Phenomenon!

The Plasti Kong Phenomenon:

There are multiple elements to this unique public awareness project: 1) Illustrated Sci-Fi Novel, 2) Mobile "Monster Hunter" Game App, 3) Public Service Announcement (PSA), 4) Plastic Pollution Solutions Guide, 5) Plasti Kong merchandise, and 6) The website (www.PlastiKong.com), which is the portal to all of it.

It starts with a visual story: The Rise of Plasti Kong. Here's a summary:

In a world drowning in plastic waste, an aging scientist attempts to create a biodegradable form of plastic. But, instead, a freak lab accident creates a mutant hybrid polymer monstrosity. Then, an unusual chain reaction activates a worldwide army of dangerous microplastic menaces. They go on a rampage against humanity and must be stopped! But how? We must try every weapon available to defeat them in landfills, recycling centers, floating garbage patches, and the streets of our cities. Thanks to a super, fast-acting enzyme (that eats plastic) created by Indigenous scientists, the solution is at hand. But will it be enough to end the reign of terror begun by Plasti Kong, King of the Polymer Giants?

Read *The Rise of Plasti Kong* to find out. Play **The Rise of Plasti Kong** Game to join in the battle! Go to www.PlastiKong.com to watch the "Mother Earth is Drowning in Plastic" PSA, purchase Plasti Kong merch, and donate to support organizations dedicated to ending plastic pollution. Learn how you can do more to help end plastic pollution and save the inhabitants of planet Earth from the MONSTER that is plastic!

For more information, contact:
Gary Robinson / Imaginative Content / 805-245-9630
P.O. Box 1123 / Santa Ynez, CA 93460 / USA

® *Trademark registration for Plasti Kong pending.*
*The Word Genie A.I. app assisted in the creation of the text of this guide.

www.ingramcontent.com/pod-product-compliance
Lightning Source LLC
Chambersburg PA
CBHW080552030426
42337CB00024B/4851